How to catch a mole

By

Armour Roberts

Copyright © Armour Roberts 2013
Photographs © Armour Roberts 2013
The Moral right of the author has been asserted.
All rights reserved. No part of this book may be reproduced in any form other than that in which it was purchased and without the written permission of the author. This book is licensed for your personal enjoyment only. This book may not be re-sold or given away to other people. If you would like to share this book with another person, please purchase an additional copy for each recipient. Thank you for respecting the hard work of this author.
www.armourroberts.blogspot.co.uk

2013

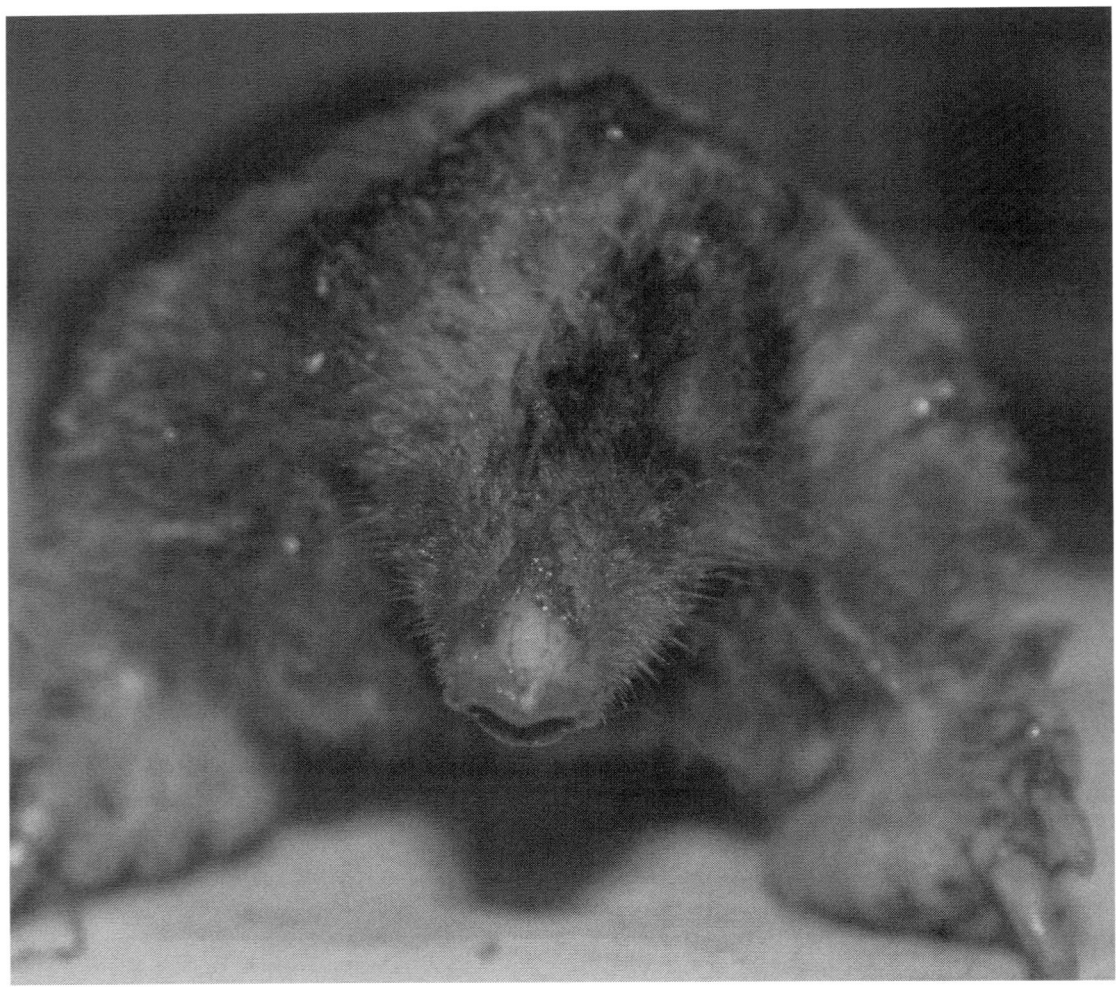

The Mole

The mole is a beautiful, hardworking, intelligent mammal, which can quickly become your worst enemy. No amount of windmills, mothballs, vibrating sticks and upturned bottles will deter a mole once it has moved into your prized lawn.

Dedication

This book is dedicated to my good friends James and Jo, also to my long suffering wife Becky, without their help and knowledge my mole catching manual would never have come to life.
.
My faithful Accomplice Ruby.

Table of Contents

Introduction
Why get rid of a mole?
Why is mole there?
The molecatcher's Tool Kit
What are 'runs'?
How to probe for a run
Where to set the traps
How to place the traps into the runs
What if I uncover a Junction?
Fine tuning your traps
Afterword

Introduction

My name is Armour Roberts and I grew up in a small welsh village in North Wales. I have been a professional molecatcher for over 25 years and have trapped moles in every type of terrain and soil type imaginable, from a 1000 acre hill farm in the depths of an extreme welsh winter to a small village bowling green in the heights of summer. I have learnt the hard way that the weather and different soil types play a major part in the types of traps you should use and where they should be placed in order to secure an effective capture.

In this book I am going to share with you my secrets and useful tips which I have used to great success over a quarter of a century. How do I find a run? What type of traps do I use? How do I set the trap? Why has a mole invaded my garden? I will answer these questions and many more.

I use only two types of traps, so they are the only two I will be instructing on in this book. They are the Duffus Barrel Trap and the Talpex Claw Trap they are humane kill traps. I have illustrations and instructions on not only setting the traps but also little tips on how to fine tune them to double your success rate. I believe that with the right equipment and a little knowledge anybody can catch a mole.

So if you're a farmer, greenkeeper, gardener thinking of becoming a professional molecatcher or just want to get rid of the mole invading your garden/yard this is the book for you.

Warning

This book contains images of dispatched moles. I felt it was important to include these images as there are right and wrong ways the mole should be caught to ensure a quick kill and minimise suffering.

Why get rid of a mole?

There are many times people have said, "How can you kill such a lovely creature"? This statement is correct; moles are incredible mammals and should be admired for there persistent excavation of tunnel systems, moving tons of earth during their lifetime, but here lies the problem. Moles remove tons of earth from the ground creating tunnels that can collapse and cause many injuries such as broken ankles, this can mean expensive insurance claims especially on places like golf courses and residential homes. Mole runs on paddocks can cause a problem too for young foals and livestock, they need to be destroyed if they sustain a bad enough break, or the farmer receives a costly vet's bill. I get called out a lot in early spring to do acres of hay fields, if the mole hills are left, in the harvesting season they can't be seen and are harvested with the hay. This can result in the hay having to be sold cheaply as it is full of dust and stones. Also, the damage caused to the machinery can run into thousands. The soil contains the Listeria Bacteria which is present in the mole hill and can be transmitted to the hay.

I am solely a molecatcher who uses traps and have never used any type of poisonous gas or toxic solution. This is because it is often a nasty painful end for a mole and you have no evidence you have eradicated the mole.

Up to 2006 strychnine was the main source of mole control. A man with a bucket of earthworms dipped in it would make a hole in the tunnel system and drop the worm into the run then plug up the hole. The mole would eat the worm and then die a lingering painful death underground. Apparently the poison could lay active in the ground for years and the risk of secondary poisoning to none target species like birds of prey and dogs was huge. Eventually in 2006 for the reasons I have stated, strychnine was outlawed and the mole population flourished.

People say it's impossible to clear large areas of moles without using gas or toxins but I disagree and have cleared hundreds of acres of land using traps alone. The benefits about using a trap is that you have an end result to show your customer and you know exactly which area you have caught the mole in, so you can knock the top off the mole hills to see if any more activity occurs there.

Why is mole there?

Before setting and laying traps you need to understand a bit about what the mole is doing, digging up your beautiful landscaped lawn and why he is there…
It's all to do with his pursuit of worms and grubs, moles don't eat plant roots or bulbs they are carnivores. His mole hill is created when he is clearing his tunnel system of debris and soil. He pushes it forward until he can no longer move it, then he pushes it up a vertical tunnel out of the ground creating a mole hill. With every bit of movement above his tunnel system eg: mowing the grass or playing football, more soil drops into his tunnel system that needs clearing, thus creating more mole hills or just adding to ones already created. Worms will fall from the turf roots into the tunnel system or move up into it from the deeper ground and the mole moves around hoovering them up. Once a mole is there feeding and you've got mole hills popping up everywhere, the only way he'll go on his own is if the worms disappear or he is killed by a predator eg: an owl or a weasel. Sometimes people get a couple of mole hills then nothing, this is usually due to there not being a sufficient food source so the mole moves elsewhere. But mark my words no amount of mothballs, windmills and vibrating sticks will get rid of him if the mole has food. He will continue making bigger tunnel systems to trap more worms in, therefore making more tunnels and mole hills in the process. This is when you get a professional molecatcher or after reading this book attempt to catch the mole yourself.

The molecatcher's Tool Kit

Traps

I use two types of traps and find them to be the best mole traps you can buy.

Trap 1: The Duffus/Barrel Trap

Trap components 1: Trigger pins, 2: Catching loops, 3: Top plate, 4: Trigger loops.

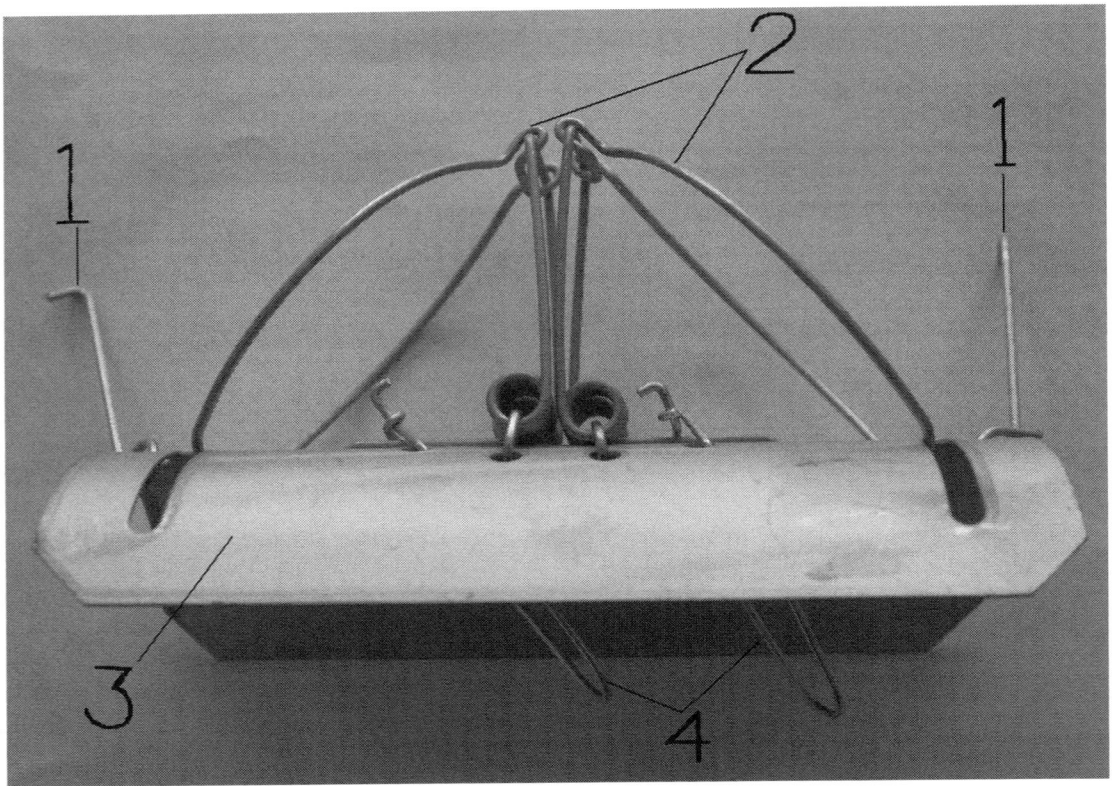

This trap consists of a curved top plate which recreates the roof of the moles tunnel system. It is set by pushing down the catching loop and swinging a trigger pin over the top which is then hooked onto the trigger loop (see pictures 1-4). This process is then repeated on the other end of the trap, meaning you can catch the mole in either end or catch two moles in one trap.

Picture 1 setting a Duffus Trap.
Push down top springs through top plate.

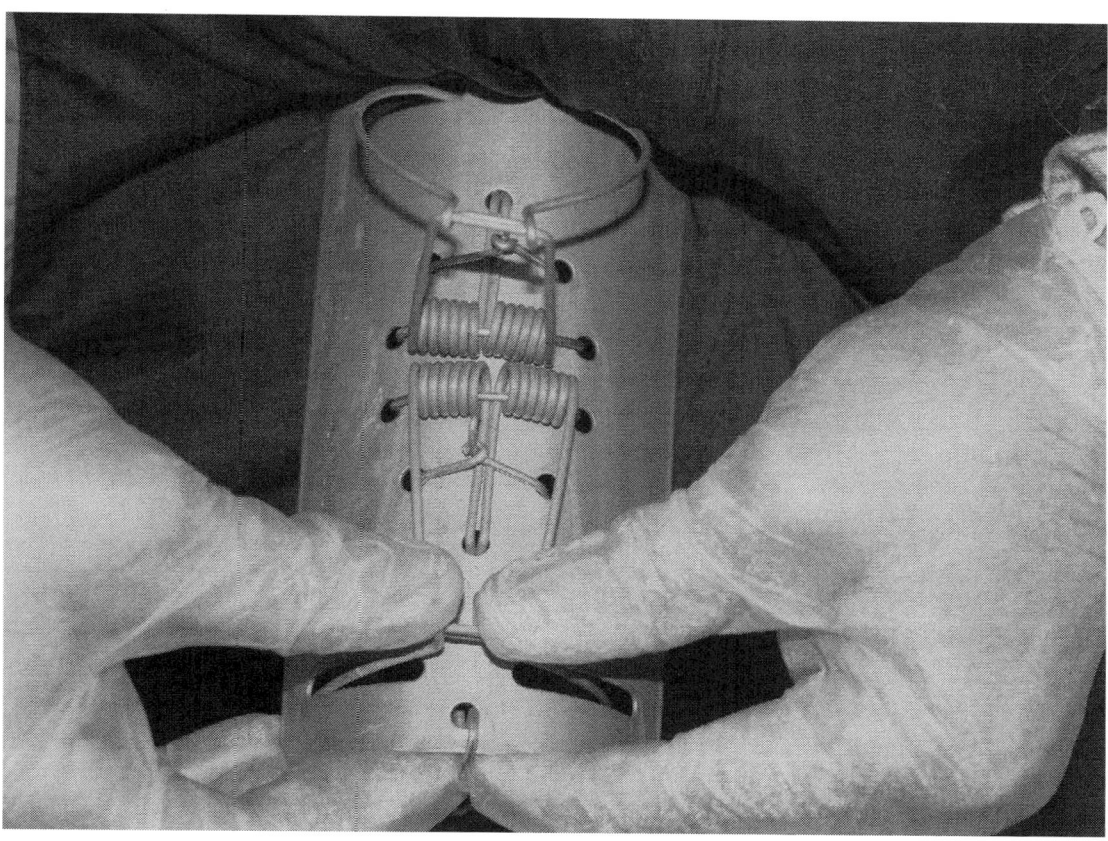

Picture 2
Whilst holding down top spring, swing over the trigger pin.

Picture 3
Hook the trigger pin onto the top of the trigger loop.

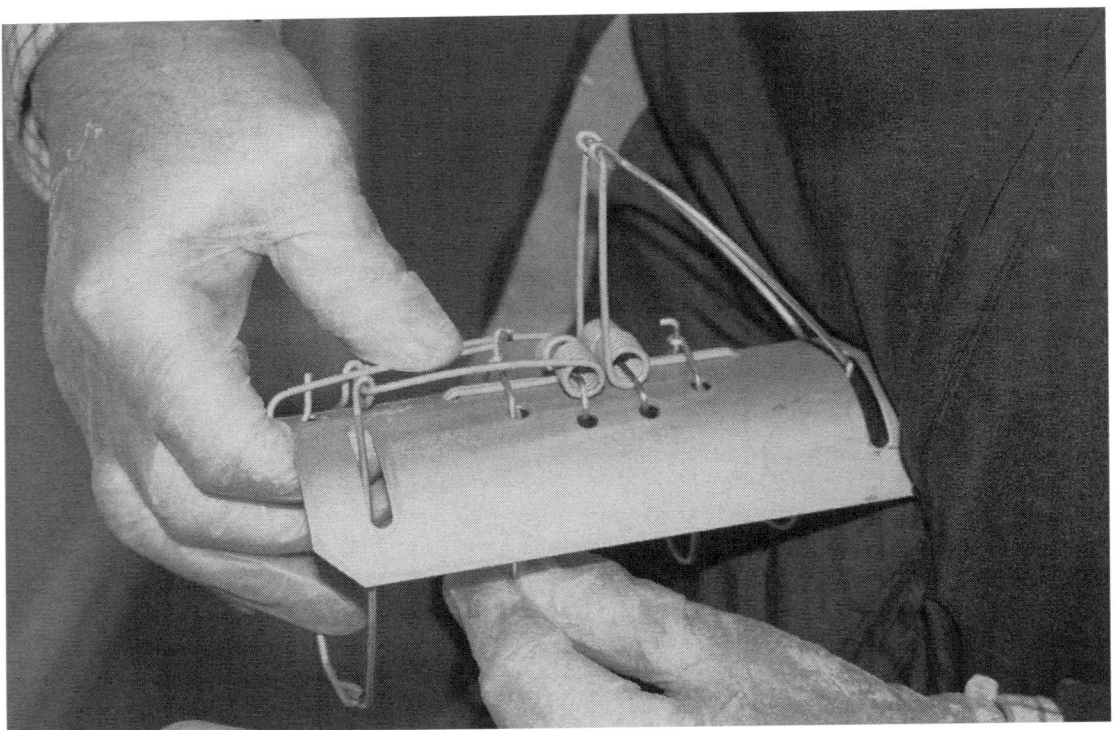

Picture 4
Then repeat this process on the other side.

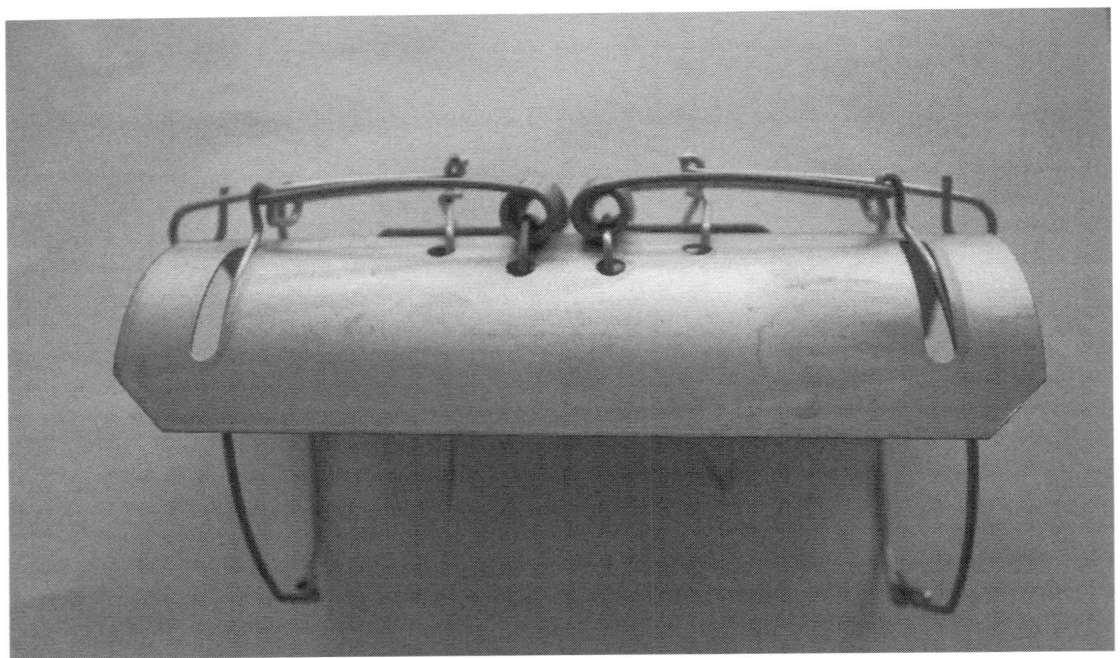

Picture 5

The best features of the Duffus Trap are that you can catch two moles at once and it is easier to backfill to exclude any light, due to its curved top plate recreating the tunnel roof; this is a hard trap to beat.

Every professional mole catcher I know has, from time to time, returned to a Duffus Trap to find an injured mole caught in it. Any mole catcher that says they haven't is either inexperienced or being untruthful, however, this is not common and this is still regarded as the best trap by a lot of professionals. I, however, rate my next trap along side it, and always use a mixture of the two together, it is the Talpex Claw Trap. I have never returned to this trap to find an injured mole in it. Every mole I have ever caught in one has been dead and the result of a quick kill. Both traps have different qualities. For instance, the Claw Trap can be harder to keep the light out when it's backfilled, whereas the curved top plate of a Duffus trap gives it a ready made roof. Also a double catch can be achieved, but the Duffus is more fiddly to set and can get blocked with soil; this is where the claw trap comes in. If you buy British made quality traps you will find these two types of traps are all you will need to be a successful molecatcher.

What to do if you find an injured mole in a trap is covered later, in the fine tuning your traps section.

Trap 2: The Talpex/Claw Trap

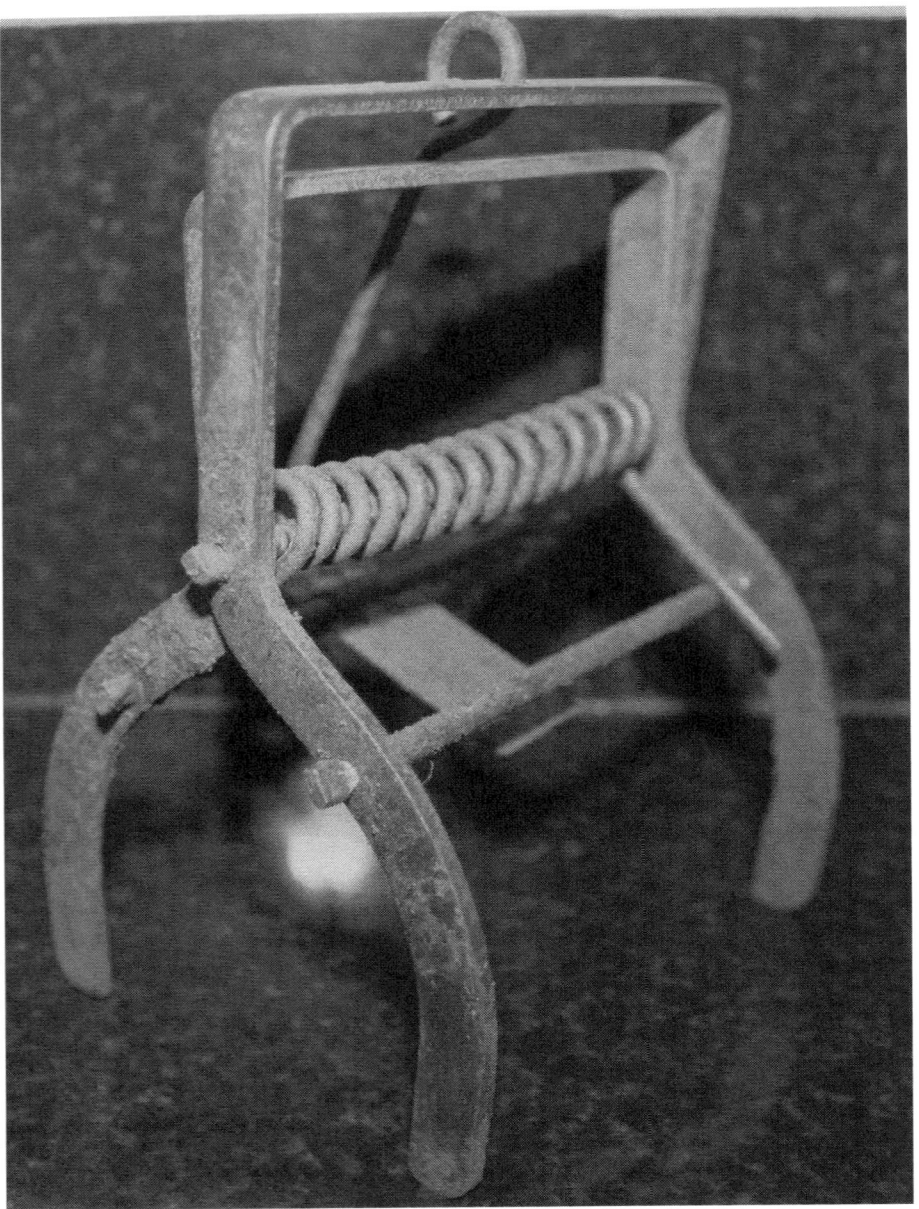

This trap consists of two handles that are squeezed together to open the catching claws. Then a pin is brought over the top of the handles and placed through a hole in the trigger plate (See pictures 7- 10). The trap is triggered by the mole moving earth forward through his tunnel system and into the trigger plate, springing the trap. This is a great trap for a mole that keeps filling a Duffus trap with soil and the best trap for trapping a mole in its feeding runs.
These traps do require quite a squeeze to open the jaws and some people may struggle with this, so you can also buy a pair of claw trap setters, basically two pieces of metal that elongate the handles lowering the pressure required to set the trap.
Also the Talpa trap is a claw trap that has fixed elongated handles that makes them easier to set, but the handles can be a hindrance as they stick up out of the ground quite far and can easily be knocked and are quite hard to cover to exclude any light.

Picture 6

Front view of a set Claw Trap. The trigger plate is tilted downwards into the run.

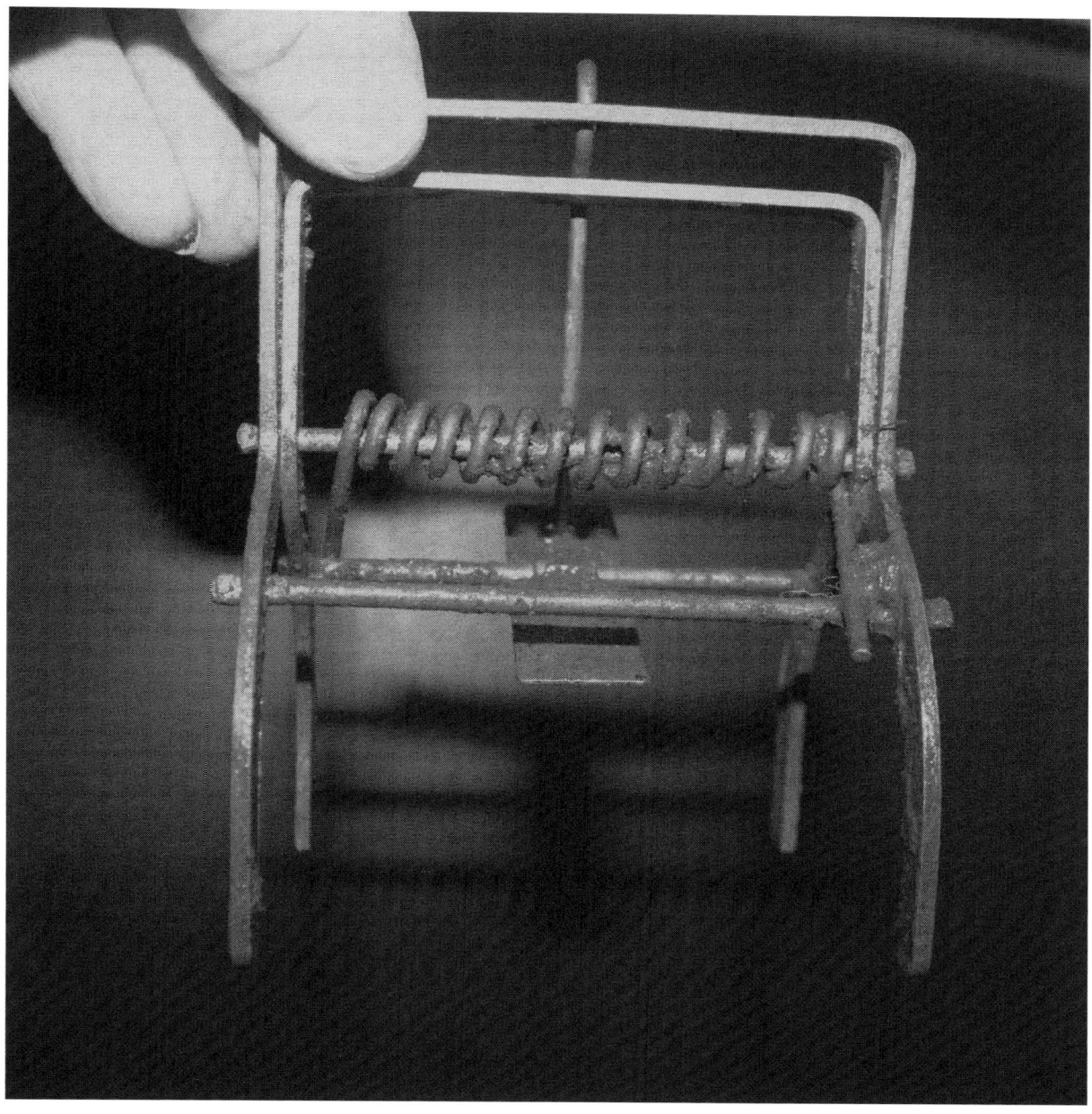

Both sets of traps need little maintenance; mine have never been oiled or required any special treatment. The only things I do before setting them, is to work the springs a few times to remove any old soil and maybe give them a tap with my trowel. Some of my traps are years old and still going strong but over time the springs do become weaker. If I am concerned about any traps strength I will set it on the floor and trigger it with the handle of my tunnel probe, if it's easy to remove and doesn't offer any resistance I will take off the trigger pins and keep them as spares then throw the rest into the bin, I will then replace them with new British made traps, or buy the genuine Talpex trap and find a reputable Duffus trap supplier.

Picture 7 setting the claw trap.
Squeeze the handles together to open the catching claws making sure the pin is flipped over the handles.

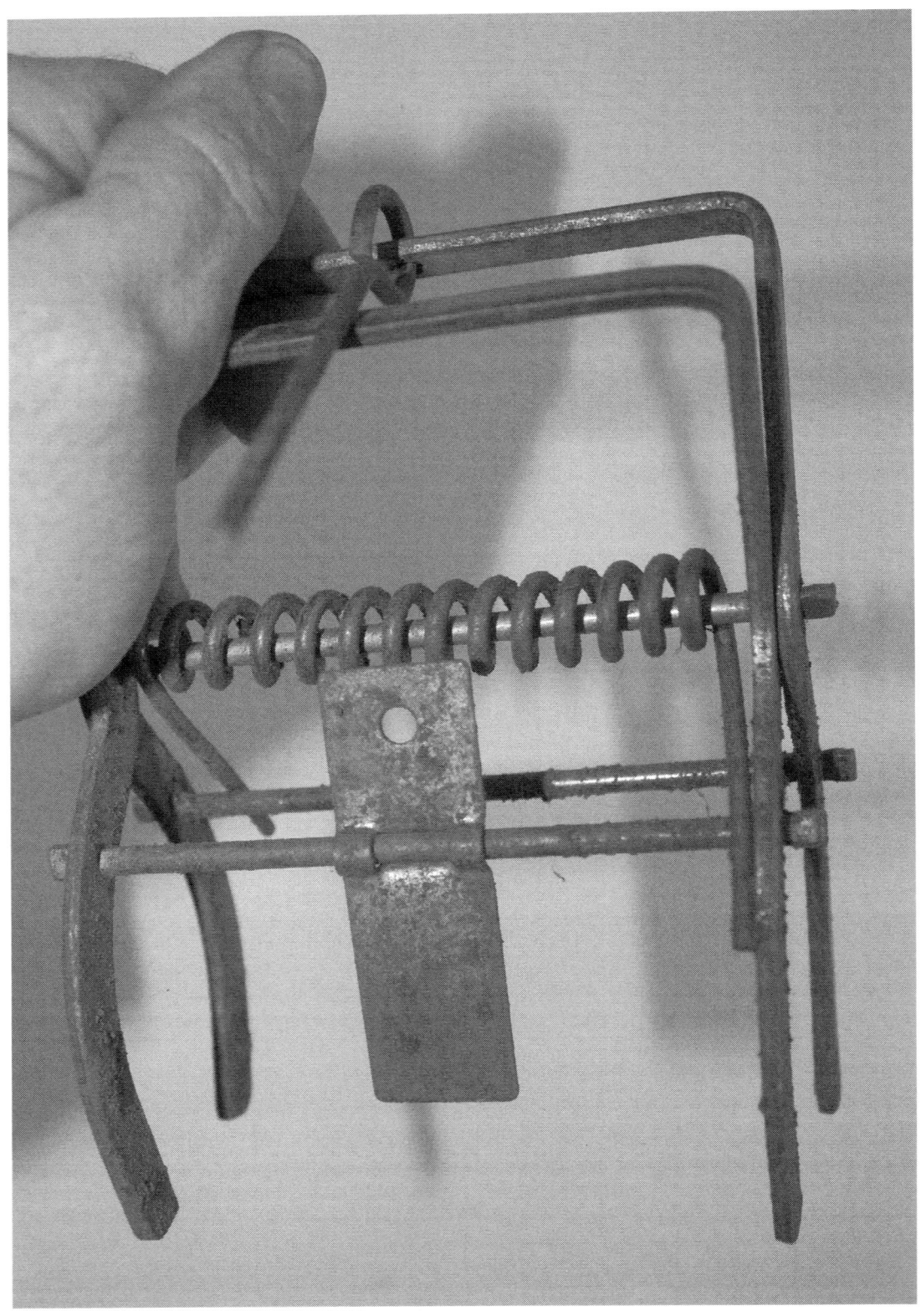

Picture 8

Place the pin through the hole in the trigger plate.

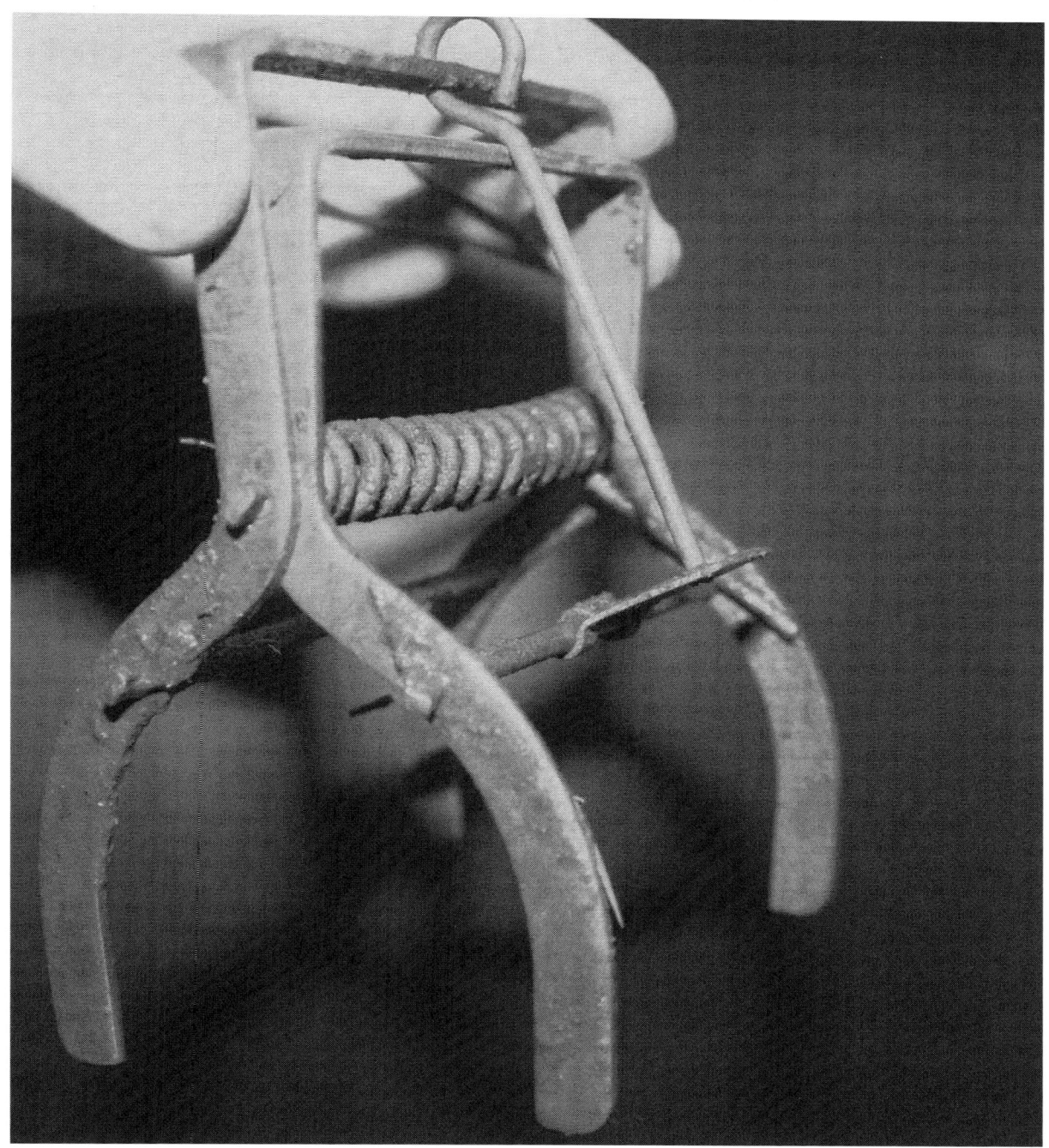

Picture 9

This is a wrongly set Claw Trap, the trigger pin is pushed too far through the trigger plate and the trigger plate is not central in the trap.

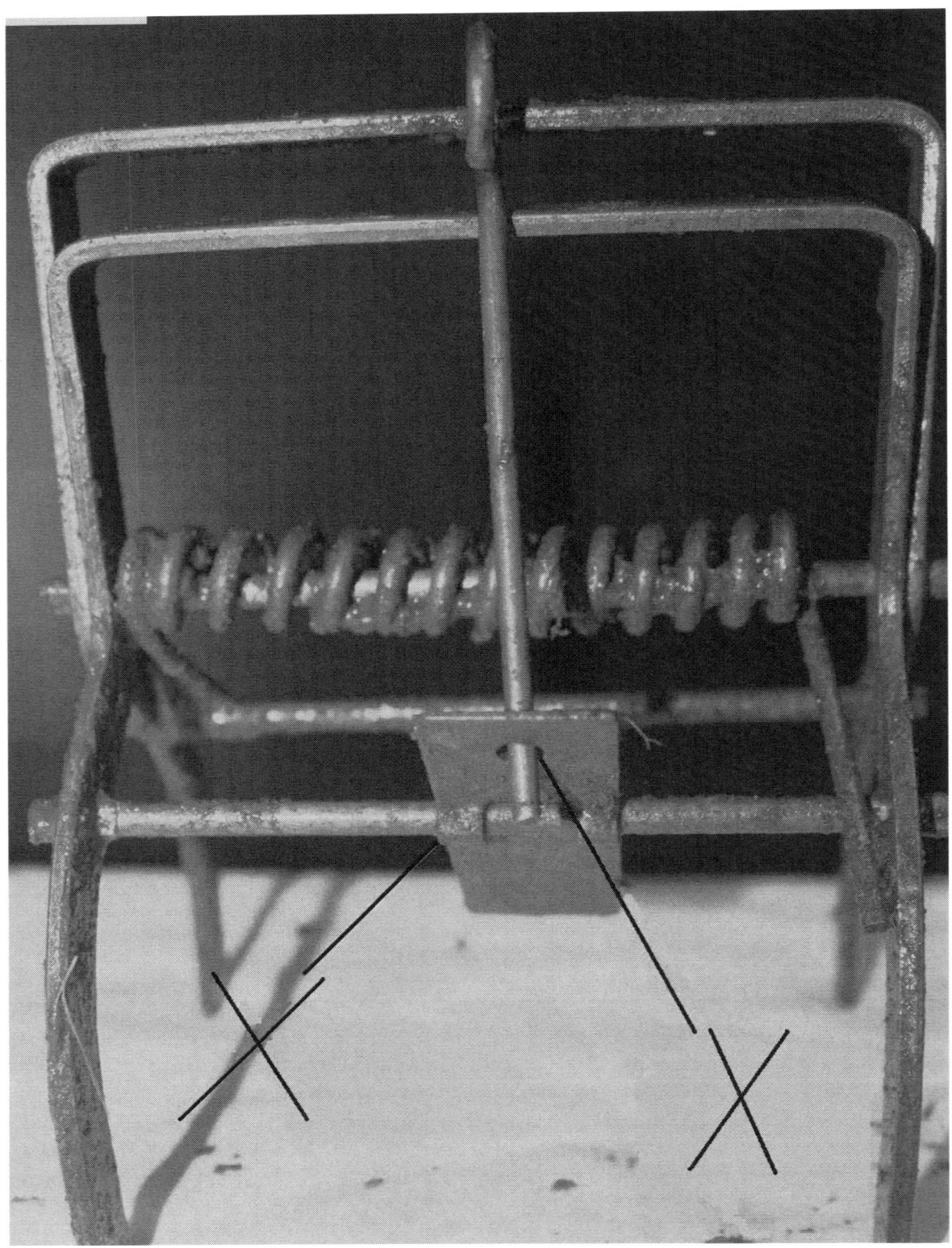

Picture 10
This is the correct way to set the Claw Trap. There is none of the trigger pin sticking through the trigger plate, also the trigger plate is now in the centre of the trap.

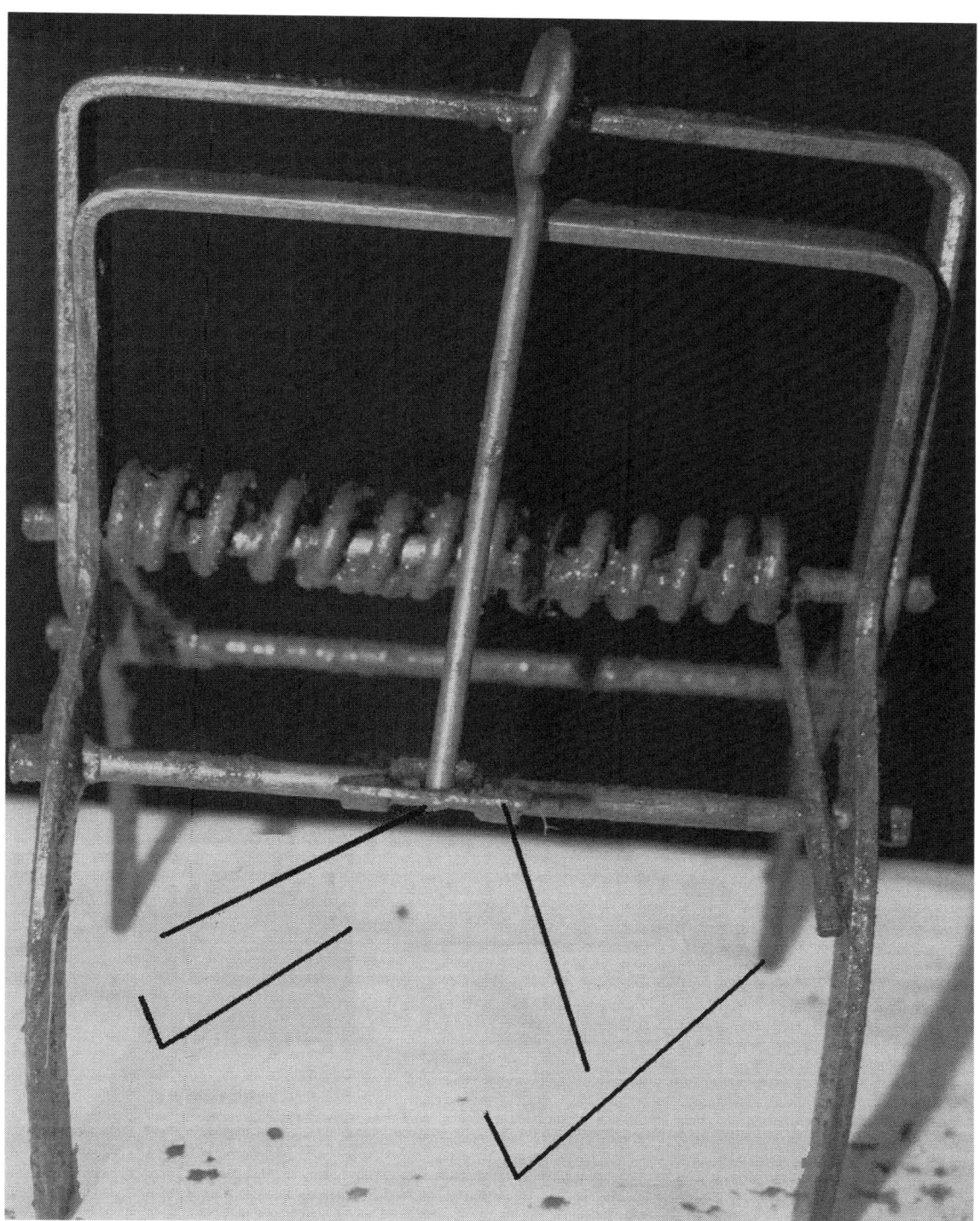

Picture 11

A mole caught correctly. Note: If the trigger plate was not central when set the mole would have been caught either too far down his body or may have been injured and crawled off in agony. Remember the mole is your quarry and must be respected, so to just catch a mole won't do you have to do it properly.

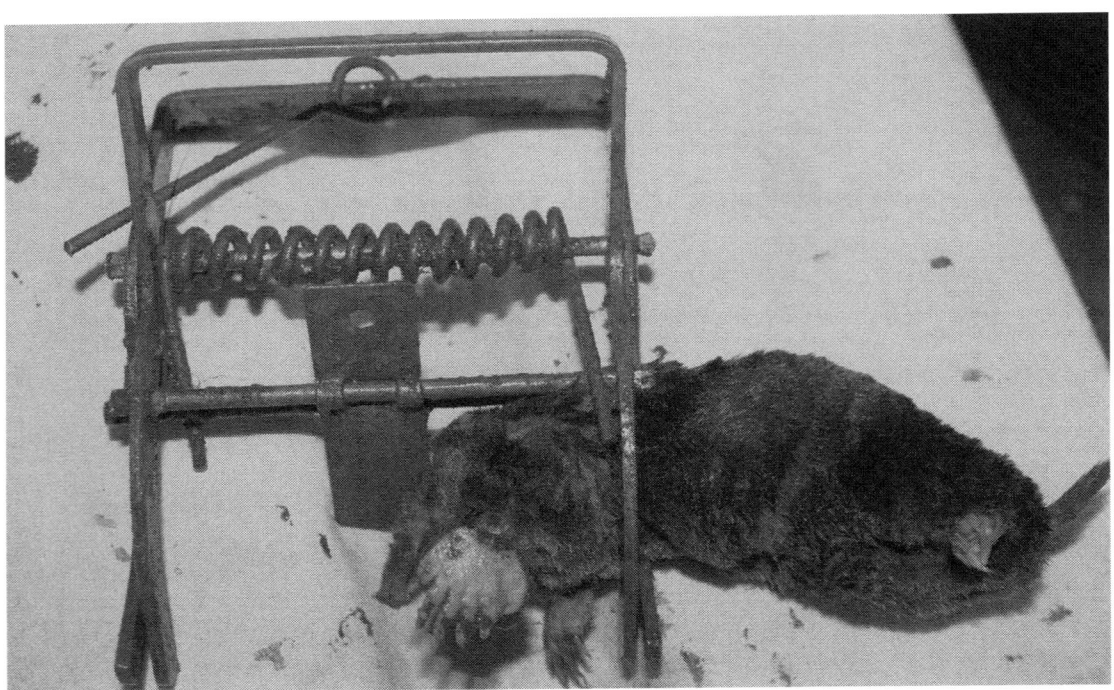

I must stress there are a lot of cheap copies out there, which are substandard. Remember no matter how frustrated you are about your mole problem, if you are to consider trapping it must be done in a way that will mean the minimum suffering to the mole and a quick and efficient kill is achieved. You will not get this by buying cheap traps. The market is flooded with cheap copies on which the springs are weak and the metal used is bottom grade. All the traps I use are British made. Expect to pay at least £5.00 per Duffus Trap and £8.00 per Claw Trap, if you see them cheaper beware. Cheap traps can not only injure or maim the mole but can be the sole reason you are not having any success. Good quality traps are usually stamped with the makers brand on the trigger plate on the Claw Trap and on the side of the top plate on the Duffus Trap. There are many other traps that are gaining good recognition in the molecatching world, the Nomol and the Trapline Traps especially. I have colleagues who now use these and I've heard good reports but it would be wrong of me to instruct on them as I have never felt the need to deviate from the Duffus and the Claw Traps. I feel these two traps used together are sufficient in any molecatching scenario.

Although the Duffus Trap and Claw Trap aim to achieve the same end result eg: a correctly trapped quickly dispatched mole, the principle behind them is a bit different. When you start to understand this you will quickly learn which type of trap you should be using in each scenario.

The Duffus trap is placed into the moles tunnel system as a recreation of the moles tunnel and relies on the mole travelling into the trigger loops at speed or by giving them a push to see if he can move them, this is what a mole would do naturally if it encountered a blockage in its run, this is why the Duffus is best set into the main runs or the gateways into the feeding area. The Claw trap, however, is placed into the moles tunnel system to recreate a blocked run and is triggered by the mole trying to clear the debris by pushing the soil forward into the trigger plate. This is why the Claw trap is a good trap for placing into the feeding area, as the mole is always moving soil in these runs and would expect to come across a blocked run. If you get a mole that has blocked your Duffus remove it and set a claw trap into the same hole, 9 times out of 10 you'll catch the mole. I will instruct on how to set the traps into the runs later.

Picture 12

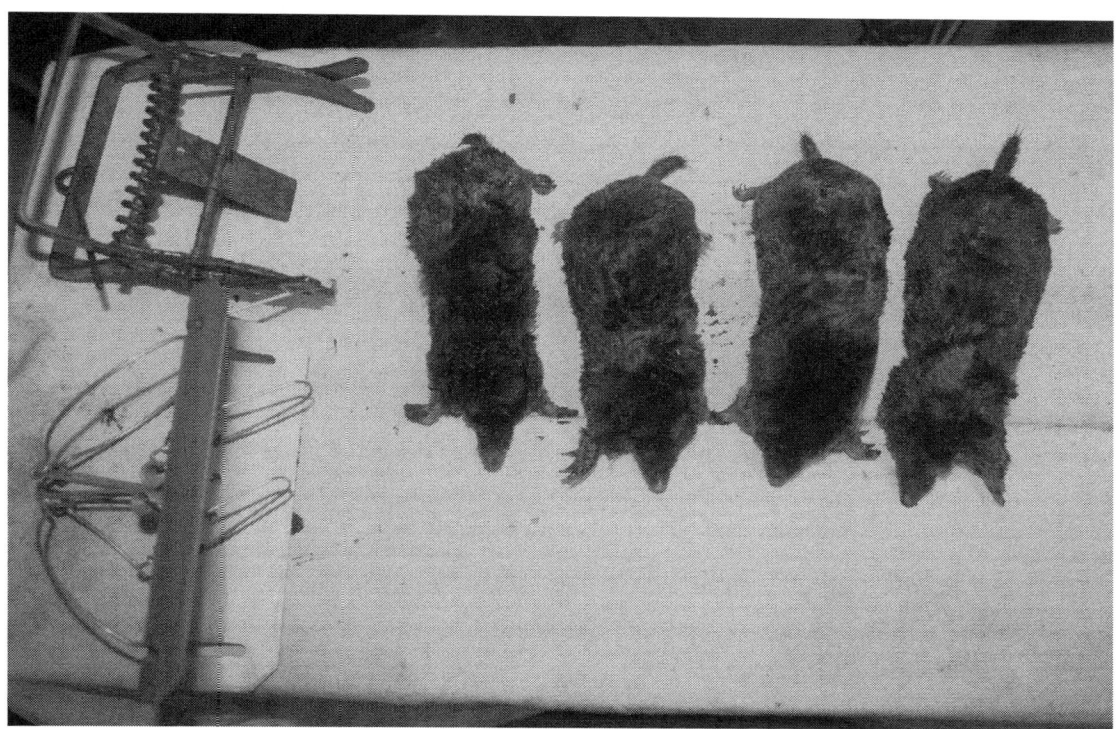

My Tunnel Probe

Picture 13

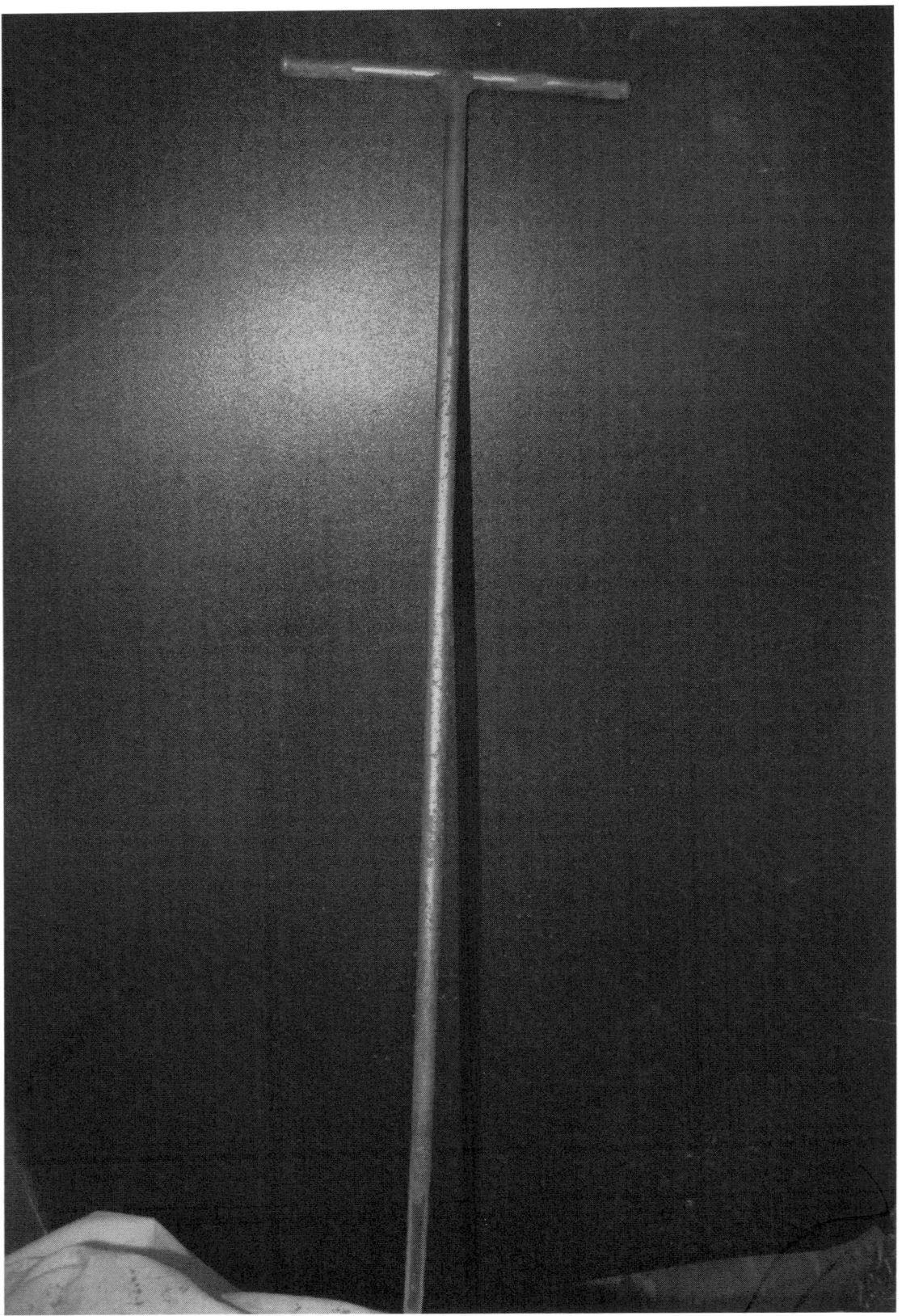

Picture 14

The tip of the probe.

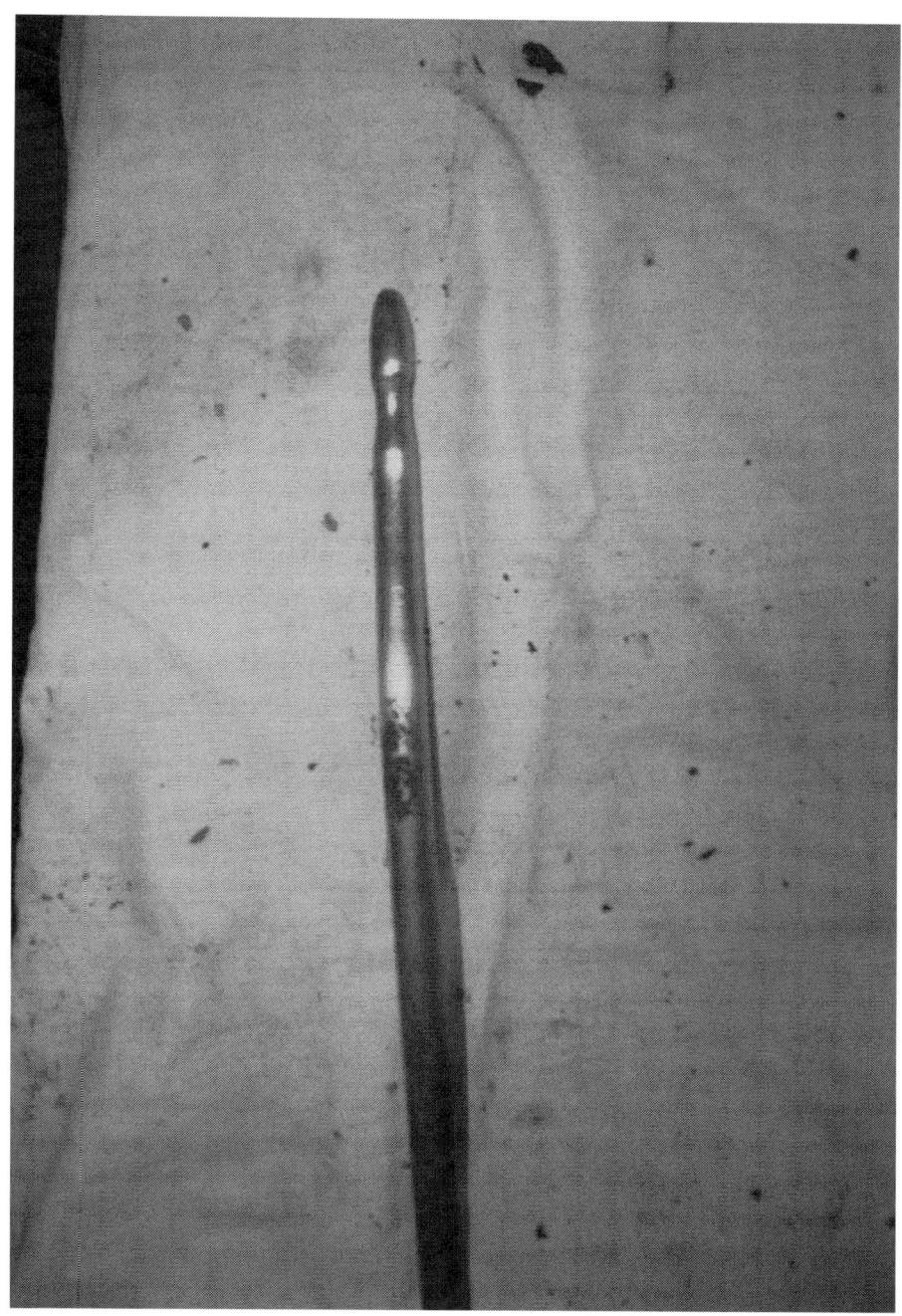

You can tell this has had a lot of use, and can see the average depths of the runs I am setting my traps into, this is shown by the shiny section of the probe tip, which is about 7 inches.

Picture 15

The T shaped handle is used for firming down the base of the run

I've owned this probe for 15 years and it was made by a friend of mine that was a welder. It is basically an 8 inch steel bar to form the handle and a 3ft length of steel to form a T shape and the very tip being of a bigger gauge. This is what I use for finding the run before setting my trap. The principle of the tip being a bigger gauge is that when it's pushed through the ground into a tunnel you will feel the probe drop into the run and hit the tunnel base.
If you haven't got a probe a normal short length of bamboo garden cane will make a good temporary tunnel probe, but remember you will also need something to flatten down the base of the run because this is an important process in setting your traps.

Gloves

Picture 16

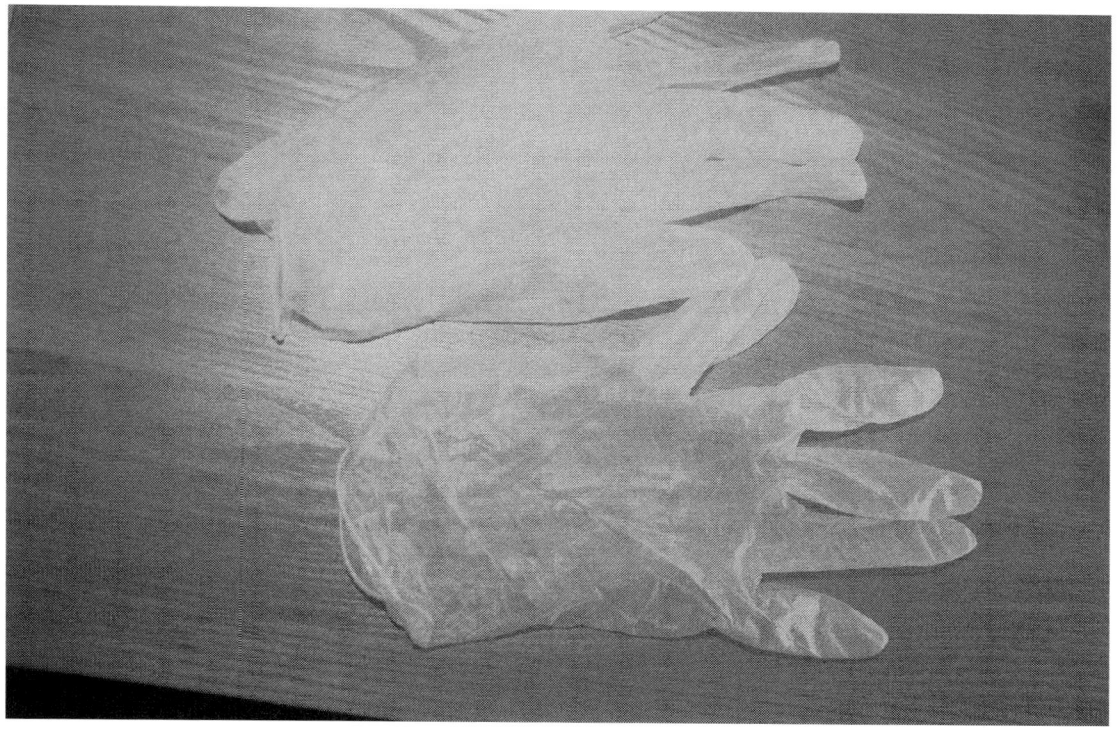

I use thin vinyl gloves when I'm mole catching they're readily available and around £4.00 per 100. I use these simply to keep my hands clean and so I don't pick up any nasty bacteria from the moles or soil. The latex version has a strong scent so I avoid these. Also you still have good sensitivity for setting traps, as they are so thin. Some people think it's to do with the mole not smelling your scent but it's not.

Once I called to a house where the gardener had hooked up a hose to the old central heating tank and flooded the runs with kerosene this was before it was worth more than gold! I had been called because the moles were still there, Kerosene smells like diesel yet the moles were still actively working in these runs. I caught 3 that were dripping wet in it. So I don't think a moles sense of smell is as good as some people think.

My Trowel

Picture 17

Lots of molecatchers differ on what type of tool they use to cut out the turf to uncover the run. My choice is simply an old brickies trowel that has been ground to a sharp V shape. This is ideal for taking out perfect sods of turf or earth. The sharp sides cut like knives and this enables you to slot the turf back into the ground when you have removed the trap and are repairing the trap site. You wouldn't even know there had been a trap in situe, which is ideal for golf courses or fine turfed areas. Also the sharp sides enable you to slice through tree roots to remove them from your trap site.

My Kneeling Pad and Bag

Mole catchers spend a lot of time on their knees setting traps, so a kneeling pad or knee pads are essential. I prefer the kneeling pad as you can just throw it on the ground whereas with knee pads you're forever adjusting the elastic or pulling them up.
Over the years I've owned lots of different bags or rucksacks. The one I use now is just an old council recycling bag. It's lightweight, waterproof and capable of holding 50 traps, however, if you are going to be climbing over styles and fences all day anything becomes heavy. If you are working on farms or golf courses it is always worth asking if you can borrow a Quad bike or similar to help you get around quicker.

Markers

Picture 18

Picture 19

My trap markers consist of two different lengthened galvanised pins. The long one I use for large paddocks and long grass and the shorter one for fine turfed areas and private gardens. I always tie a new piece of red and white warning tape through the tops to make them visible to me when I return to check them. It also acts as a warning to children and pet owners. However, sometimes you need to be discreet and find other methods of marking them that only you know about.

What are 'runs'?

When people refer to the moles run they are actually talking about the tunnel system in which the mole is actively working. They are an underground network of tunnels in which the mole travels, and periodically has to remove soil or debris.
There are different types of runs,
1. The ones linking the mole hills are feeding runs and the mole is busy in these moving earth and eating worms and grubs. These are usually about 6 to 8 inches down but this can vary on soil type. Some molecatcher's say you can't catch in the middle of the feeding run and you will get a trap full of soil. This can be true. However, it all depends on your soil type and type of trap you use. I would say majority of my catches are in the feeding runs. I believe that with the right traps and a little knowledge anybody can catch a mole.
2. There are the travelling runs, these are what people refer to as main runs and they are the tunnels the mole uses to gain access to the feeding runs. These are normally the same depth as the feeding runs. These runs are the best choice of tunnel to trap but I have just as much success trapping in feeding runs. The travelling runs are usually located around the borders of wherever you are trapping, so for instance in a garden bordered by fence panels or hedges probe at the base of the fence panels about 6 inches out or under the hedges, also along the edges of a path or patio. I do a lot of paddocks and fields and I always find very productive runs underneath the bottom fence rails around the perimeter. This is where your Duffus Trap will come into its own and set correctly into these runs it should start producing lots of doubles eg: Two moles in one trap, depending on how many moles are present.
3. There is also a secondary tunnel system. This is a deeper tunnel system which can go down to 3ft deep. This is used by the mole when the weather drops below freezing or there is a hot summer or drought. The worms follow the moisture down, therefore, the mole follows. These tunnels are a lot trickier to trap in than the other two so I wouldn't recommend trying. It is very labour intensive as a deep hole is required, which means more earth needs excavating and usually they are very hard for a novice to find due to their depth. It is these runs in which the mole spends a lot of time, in freezing or drought conditions. This is when you may think the mole has simply disappeared as no new mole hills come up. As soon as the weather warms up or the ground gets wetter the mole hills start to pop up again due to the worms returning to the surface.

Picture 20

A picture showing a run, the tunnel is running from left to right and this one is in clay soil about 4 inches deep.

You can tell if a run is active because the sides and base will be smooth this is due to the moles feet and underside making contact with them as it travels through the tunnels. In an unused run the base and sides look dull and dry, also the turf on top of the run may have sunk, an active mole would push this back up with his back if he were working there.

There is a section later that covers what to do if you find a three way run entitled,
What if I uncover a junction?

How to probe for a run

I am constantly asked this question and the truth is, 'it's easy'. Once you've done it a few times you soon get the hang of it. I will start with the easiest tunnel for you to find, however, this is just to get you used to finding a run, and is not the tunnel I would recommend you set your traps in to begin with. Once you've found this tunnel the same principle applies to all runs and with practise you will soon be finding main runs and gateways into the feeding area. The run you should start with is in the moles feeding area i.e.: the area with all the mole hills. Firstly look for the freshest mole hills, you can tell these as the soil on them looks fresher, also in the summer months the earth brought to the surface is darker due to it being damper than the old mole hills that have dried in the sun. Any molehills that appear flat or have tufts of grass or weeds growing from them are old, but that's not to say they wouldn't have an active run underneath them.

When you have identified the most recent molehills look for two adjacent to each other, then using your probe about 4inches out from the mole hill and in line with the adjacent hill push the probe down gently into the ground. As the tip of the probe goes through the turf you should feel it suddenly drop then stop, this is the probe dropping into the run and stopping when it's hit the base. If you don't feel this, take out your probe and move sideways about 2 inches then probe again. Continue this motion moving clockwise around the molehill and eventually you will find a run. Be patient and don't give up, it does take a fair bit of practice. The mole is constantly flattening and smoothing the base of its run with its stomach and feet and gets use to how this feels, this is why before inserting the trap we remove all loose soil then tamper down the base of the run, to recreate how the mole would expect to find it, this is why its important not to probe too deeply and damage the base of the run. Moles are sensitive to anything that feels unusual or out of place. I use the handle of my probe to do this and will usually insert the base inside the run too.

Where to set the traps

Take a look at the perimeter of the area where the mole is working. Is there a fence or hedge as a boundary? Is there a paddock or some woods next door and are there mole hills in those areas too? Then look for a line of mole hills going back to the boundary of the area. This could be the moles entrance into the feeding area also known as a gateway.
If you find this line of mole hills, follow them back towards the perimeter then probe around the very first mole hill going towards the perimeter. You should find a run here and this is an ideal place for a Duffus Trap. This is because if there is a mole working in the feeding area and another coming into it from a main run you could end up with a double catch.
Next, I would probe around the perimeter of the area looking for main runs. These are normally in straight lines under hedges or about 3inches out from the base of the fence panels. You can also find them running next to slabbed pathways and patios. These are also excellent runs to lay your Duffus Trap in because of catching a mole in either end of the trap. If the area is surrounded by a rail and post fence system most of the time you will find a very productive run underneath the bottom rails. If you can find this run, set Duffus Traps on each side of the area underneath the fence rail and if you can't find a main run don't worry for now. Go back into the feeding area where the freshest mole hills are and probe around one just in from the gateway you found earlier. When you have found the run set a Claw Trap into it and then another Claw Trap in the feeding run inbetween two fresh linking mole hills. Then finally, find the last fresh mole hill in the line and probe beyond it, not back into the feeding area and lay a Duffus Trap into this run.
The scenario I have given you is for an average garden mole problem using 5 traps. This is the minimum number of traps I will set to catch a mole in an area and I will always use a mixture of the two trap types. However, if the soil type is extremely sandy I would use more Claw Traps than Duffus. This is because in sandy soil the runs get blocked and require clearing out more often, meaning there would be more chance of a Duffus Trap becoming blocked. Therefore, this is an ideal situation for a Claw Trap. In clay soils however, Duffus traps are more productive as the runs require less clearing out by the mole. For instance, on golf courses I use mainly Claw Traps because the soil is more sand based and cricket squares I will use a Duffus Trap because the soil tends to be more clay based. This is something you will only learn with experience and in time you will be able to determine which of the two traps to use more of by simply assessing the mole hill to see what type of soil the ground has. But always use a mixture of the two types of traps to maximise your catching potential.

Picture 21

This shows my trap positions as described. The markers showing a number 1 are Duffus Traps; one is in the main run by the fence and one in the gateway to the feeding area. The markers showing a number 2 are my claw traps, set in the feeding area. I have two more Duffus Traps set after the last mole hills; this is where the mole is extending his feeding area.

How to place the traps into the runs

The Duffus Trap

When you have probed down and successfully found a run, place the Duffus Trap on the ground next to the run and using the trap as a guide cut down opposite either end of the trap. This will give you the correct length to cut your hole. Then cut the length of the trap along both sides, remember your trap should fit tightly into the hole but try not to damage the base of the run. The cuts should be made cleanly and be made vertical and you should be able to remove a perfect sod of turf. Place the turf on the ground and remove any loose soil or stones. Feel into the runs to make sure there are no stones or debris that the mole could push into your trap blocking the catching loops. If you cut the hole too big for the trap, loose soil can fall into the tunnel either ends of the trap alerting the mole to some disturbance. If you find you've cut your hole too big, don't despair simply place some grass or leaves over the gaps before you backfill (see pictures 28 -29).

Next, using the T shape handle of the probe, firm down the base of the run, then set your Duffus Trap as described in pictures 1-4. Holding the trap by the top springs, using both hands place the trap into the hole and give it a twist to see if you can bed the trap in a bit, then twist it back so the springs are upright. Make sure the roof of the trap is in line with the roof of the run. Refill either side of the trap and firm down slightly. Then completely cover the rest of the trap with a good covering of loose soil, making sure no light at all enters the trap. Remember a mole will sense any pinprick of light and will quickly back out of the trap filling it with soil in the process. You can usually put at least 1 to 2 inches of soil on top of the trap to cover it and it will still work. It is very important to exclude all light and make sure none can enter the run; the mole is used to working in pitch black. Once backfilled I usually knock some soil off the sod of turf and place it on top of the trap, this helps keep out any wind and rain and also adds an extra bit of cover.

When returning to check your traps you should be able to tell if ones sprung simply by removing the turf sod and seeing if the top of the catching loop is visible, see pic 26.

Picture 22
A Picture showing the turf cut to size and removed and the base tampered.

Picture 23
Holding the top springs, place the trap into the run.

Picture 24
Backfill over the trap with loose soil.

Picture 25
Place the turf back onto the trap site and mark.

Picture 26

The top of the catching loop is normally visible if trap has been sprung.

Picture 27
The mole has been successfully caught.

Picture 28
This picture shows a hole cut too big, notice the gaps each end.

Picture 29
Before backfilling make sure you fill the gaps with grass or leaves.

The Claw Trap

When the run has been found, place the Claw Trap on the floor in front of you and cut a square hole into the run, roughly the same size as the Claw Trap. Remove any stones from the run. Then getting a handful of loose soil, making sure it doesn't contain any stones or debris that may jam the jaws, place half of it into each tunnel. What you are doing is blocking the moles tunnel. Don't push it up the tunnel or firm it, just place it into the holes, usually this technique is required in clay type soil or when the mole hasn't got much soil to clear out from its runs, If the soil type is sandy or there is a lot of activity above ground the mole will be constantly moving earth in the runs so you may not need to place any soil up the tunnels as it may lead to a sprung trap and no mole caught, this is why understanding different soil types is a crucial part of molecatching. Then set your Claw Trap as in pictures 7-10 and place into the run making sure the trap is central in the tunnel, normally the gap from the tip of the trigger plate to the base of the run should be about an inch, then backfill with loose soil containing no stones and fill up to the top of the springs, so only the handles are visible.

I then usually knock some of the soil off the turf sod and place it over the handles making sure they will open fully. It's always a good idea to cover the trap site with an upturned seed trap as well if you have a few. Remember make sure no light enters the trap. Mark the trap site and wait.

Once traps are set, don't be tempted to mess with them. Leave them and check them at least once every 24 hours. After 3 days if you are still unsuccessful, switch your traps around using the same holes but swap the trap types. If still no good, probe again and find new runs. It does take patience but once you've caught your first mole there will be no stopping you.

Picture 30
Once the run has been found place the Trap next to it and cut the hole.

Picture 31
Take a small amount of loose earth and place some in each tunnel.

Picture 32
Set the trap using picture 9-11, then insert into the run making sure it is central in the tunnel.

Picture 33
Backfill the hole with loose soil containing no stones.

Picture 34
Cover your trap and mark; I've used a small seed tray to keep out the light.

Picture 35
If on your return the handles on your trap are open it has been sprung.

Picture 36
The mole was caught whilst pushing forward the soil placed into its run.

Picture 37
When placing the trap into the run, make sure the back of the trigger plate where the trigger pin goes through, doesn't touch the side of the tunnel as you're pushing it down. This can slide the trigger plate back up the pin meaning the trap will require a lot more effort from the mole to trigger.

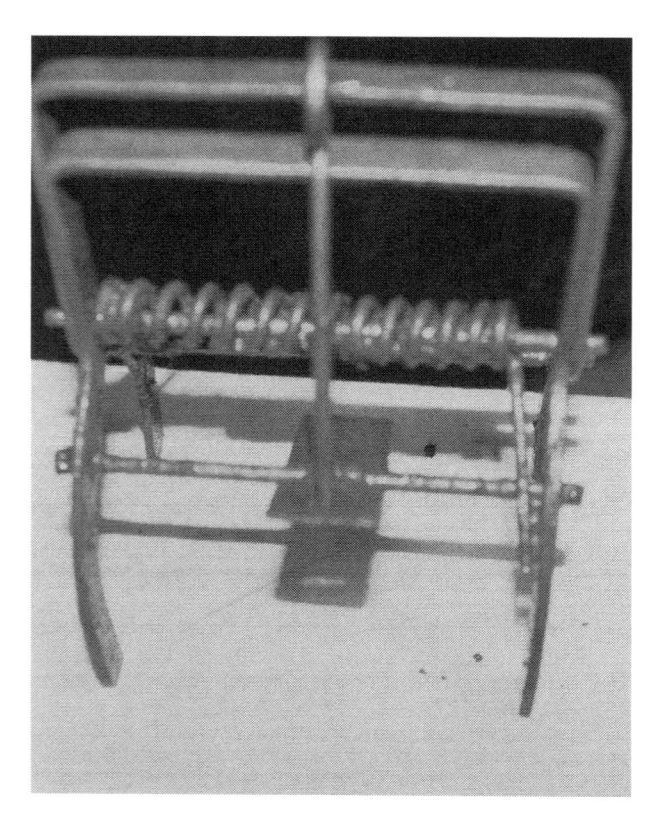

What if I uncover a Junction?

Basically a junction is a three way run, which means if you uncover one you're not entirely certain of which direction to set your trap. Usually the best thing to do is fill in the run and replace the turf and probe for a normal straight run elsewhere. However, there is one scenario where you can successfully catch moles in junctions. If you ever uncover a straight main run with a third tunnel going off into the gateway of the feeding area, set a Duffus Trap into the main run, then cut out enough soil for a claw trap to be set in front of the third tunnel, then backfill as usual. Also remember to put two markers in the ground to remind you that you have two traps in the run.

Picture 38
A junction, the main run is parallel to the slabs; the third is going to the gateway of the feeding area.

Picture 39
Tamper down main run base.

Picture 40
Place a little loose soil into the third tunnel.

Picture 41
Set both traps as earlier described, and then place the Duffus into the main run and the Claw Trap into the third run.

Picture 42
Then backfill. The soil was very dry which meant I was unable to cut the main run to size so I plugged the gaps with leaves.

Picture 43
On my return the Claw Trap had been sprung.

Picture 44
The mole was caught in the Claw Trap whilst travelling back from the feeding area to the main run.

Fine tuning your traps

There are certain things you can do to your traps to up your success rate. These things are simple steps that you perform and can mean the difference between a catch and a fail. For instance, when you first buy a Duffus Trap the hook on the top of the trigger loop is sometimes bent down too much, simply by taking a pair of pliers and bending it upwards will mean it will require less of a push from the mole to trigger. Do not bend it up too much or it will become too sensitive and keep going off with every touch or vibration. I usually bend mine just below horizontal, but every trap is different, so it takes a bit of trial and error to get it right. When handling set traps always hold either the top springs or the bottom of the top plate in the centre to avoid trapping your fingers. This is an occupational hazard though I'm afraid and I have suffered plenty of black finger nails when I began mole catching.
Sometimes, especially on the cheaper traps the trigger loop hook is quite long meaning the mole has to push really far into the trap before it triggers, this results in the mole being caught towards its rear end or the trigger loop will be too hard to push, so it will dig under your trap filling it with soil in the process. To remedy this problem the hook can be shortened with wire cutters by up to half.
When setting the trap, move the hook end of the trigger pin right to the tip of the hook on the catching loop, this is called a hair trigger and will require only the slightest touch from the mole. However, if you buy good quality British traps they are usually tuned for you.
There is however one thing I always do on any Duffus trap that I set, firstly set both sides of your trap and make sure the catching loops are vertical, then slightly bend the trigger loops towards the catching loops one end at a time. I've found by doing this most of the time I will catch the mole behind the front legs which will crush the chest meaning a quick kill is achieved. But remember only a slight bend is required, if you over do it you run the risk of the mole being caught by its foot and that's unacceptable. Unfortunately, it does happen to the best of us if you're setting thousands of traps a year. If you're ever unlucky enough to find an injured mole in your trap you must act fast for the sake of the mole. I do this with a single strike to the back of the head with my trowel. I know this sounds gruesome but it has to be done, it is your duty, because if you hadn't set the trap the mole wouldn't be injured. This is an important factor that you must consider before putting any trap into the ground.

Picture 45

A tuned Duffus trap, the small black line on the top of the trigger pin shows its old position, also the set side of the trap is hair triggered and the trigger loops have been bent slightly towards the catching loops.

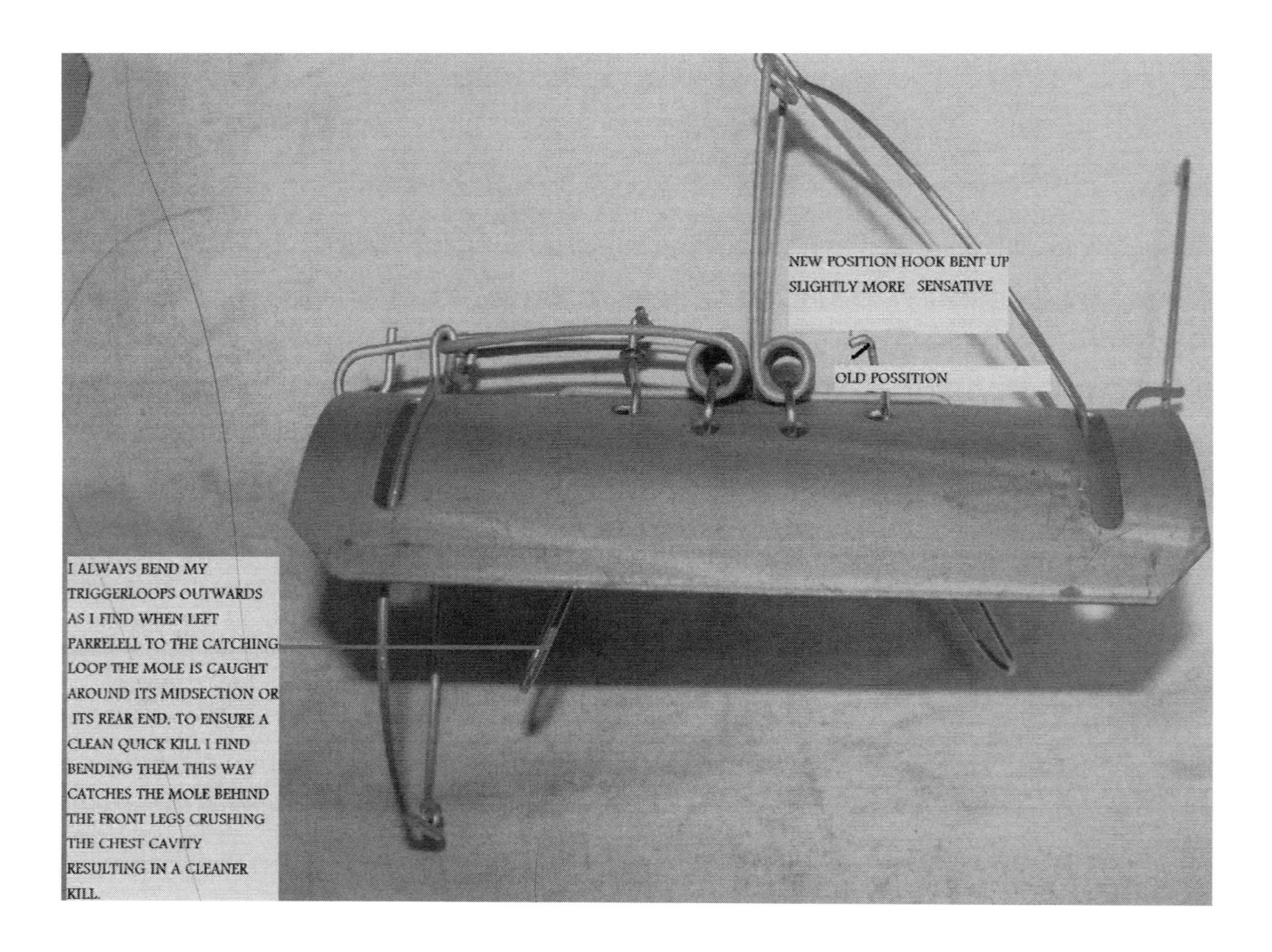

NEW POSITION HOOK BENT UP SLIGHTLY MORE SENSATIVE

OLD POSSITION

I ALWAYS BEND MY TRIGGERLOOPS OUTWARDS AS I FIND WHEN LEFT PARRELELL TO THE CATCHING LOOP THE MOLE IS CAUGHT AROUND ITS MIDSECTION OR ITS REAR END. TO ENSURE A CLEAN QUICK KILL I FIND BENDING THEM THIS WAY CATCHES THE MOLE BEHIND THE FRONT LEGS CRUSHING THE CHEST CAVITY RESULTING IN A CLEANER KILL.

A small smear of Vaseline placed on the trigger pin hooks and the catching loop hooks will make your trap more responsive.

Tuning the claw trap

To be honest there's not a lot of tampering around required with this trap, the key to its success is by making sure it is set properly as described above. I've always got a bucket of compost in the van handy because if the soil is wet clay or the grounds frosty I can backfill with the compost. The only thing I can recommend is that sometimes if the base of the run is really hard or on a tree root so the trap can't be bedded in, bend the trigger plate downwards a little.

A small amount of Vaseline smeared onto the bottom half of the trigger pin can help the trap open more efficiently.

After word

In this short manual I have been as brief as possible and have stuck to the main objective, which is to help you to catch a mole. I have passed on all of the tips and secrets I use day in day out as a mole catcher and have tried to make it as simple as possible. Mole catching is my life and has been for over 25 years and the methods I've described to you have given me a long and happy career.
So, if you are just interested in catching your own garden mole, or thinking of becoming a mole catcher, armed with this manual and a bit of patience you will soon start catching moles. If you have any difficulties understanding any of my instructions or need some extra advice, please feel free to contact me at   armourroberts@yahoo.co.uk

Happy molecatching

Armour.

THE END

Follow me on twitter. Armour Roberts moleman_
www.armourroberts.blogspot.co.uk

Printed in Great Britain
by Amazon